DATE DUE

Demco No. 62-0549

What Do You Know About
Plant Life ?

PowerKiDS
press

New York

Suzanne Slade

To Esther Hershenhorn, our fabulous SCBWI-Illinois Regional Advisor who works tirelessly helping writers grow and bloom

Published in 2008 by The Rosen Publishing Group, Inc.
29 East 21st Street, New York, NY 10010

First Edition

Editor: Amelie von Zumbusch
Book Design: Kate Laczynski

Photo Credits: Cover, pp. 1, 7–11, 13–22 Shutterstock.com; p. 5 © www.istockphoto.com/Maria Bibikova; p. 6 © www.istockphoto.com/Malcolm Romain; p. 12 © www.istockphoto.com/Hakan Pettersson.

Library of Congress Cataloging-in-Publication Data

Slade, Suzanne.
  What do you know about plant life? / Suzanne Slade. — 1st ed.
      p. cm. — (20 questions. Science)
  Includes index.
  ISBN 978-1-4042-4200-5 (library binding)
  1. Plants—Miscellanea—Juvenile literature. I. Title. II. Series.

  QK49.S58 2008
  580—dc22

                                                    2007033506

Manufactured in the United States of America

Contents

Plant Life

Plants are an important part of our world. They give us food, such as sweet fruits, healthy vegetables, and tasty nuts. Wide, leafy plants give us shade on a hot summer day. Plants also make the **oxygen** in our air. People and animals need oxygen to breathe.

Plants give animals a safe place to hide from their enemies. Many animals make their homes in trees and bushes. Flowering plants add color and beauty to gardens, parks, and yards. The many plants in our world make it a wonderful place to live.

This boy is picking strawberries. Strawberry plants have white flowers and leaves that grow in groups of three. Like blackberries and raspberries, strawberries are in the rose family.

1. What are the main parts of a plant?

Most plants have roots, a stem, and leaves.

Leaves

Stem

Roots

Young plants, like this pea plant, need their roots, stem, and leaves to grow.

2. Which part of a plant is the most important?

Every part of a plant has its own job. All these parts work together to help a plant grow. The strong stem carries water, and leaves make food. Roots hold a plant in place and take in water and **minerals**.

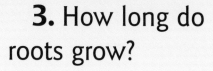

Plants have either taproots or fibrous roots. A plant with a taproot has one main root that goes straight down. Plants with fibrous roots have lots of thin roots that spread out. If you add them up, a plant's fibrous roots can be very long. A winter rye grass plant can have up to 385 miles (620 km) of fibrous roots!

If a tree is growing on steep, rocky land, you can sometimes see some of the tree's roots sticking up out of the soil.

7

4. Why are most leaves green?

The green color of leaves comes from green matter called **chlorophyll**.

While most plants have chlorophyll in their leaves, cacti, such as this one, have chlorophyll in their thick stem.

5. Why do leaves have chlorophyll?

Plants need chlorophyll for **photosynthesis**. Photosynthesis is the way plants make their own food.

6. What happens during photosynthesis?

During photosynthesis, a plant takes in sunlight, a gas from the air called **carbon dioxide**, and water. Using these three things and chlorophyll, the plant produces a food called **glucose**. Plants also give off oxygen during photosynthesis.

7. How do trees that have needles make food?

The needles on **evergreen** trees are really long, thin leaves. Needles have chlorophyll and are able to carry out photosynthesis, just like other leaves.

An evergreen's needles supply the tree with glucose throughout the cold, snowy winter. They also provide a place where small animals, like these cardinals, can be safe from the wind and snow.

9

8. Do all plants make their own food?

No. Plants called parasites use food that other plants make.

Rafflesia is a parasite from the rain forests of Southeast Asia. Rafflesia grows only on tetrastigma vines, from which the parasite draws food and water.

9. What are some parasite plants?

Parasitic plants do not need sunlight to make food, so many of these plants live in shady places. One of the most beautiful parasitic plants is called rafflesia. Every 10 years, it produces the largest flower

The clumps of leaves on this tree are mistletoe. Unlike many of the trees on which it grows, mistletoe keeps its leaves and berries throughout the winter. This supplies food for many birds.

in the world. This flower measures about 3 feet (1 m) across and weighs around 15 pounds (7 kg). Although this red flower is beautiful, it smells horrible.

Another parasite is called mistletoe. This green, leafy plant could make its own food but uses its roots to steal food, minerals, or water from other plants.

10. Where do flowers come from?

Flowers grow out of small balls called buds.

11. Why do flowers seem to suddenly appear?

Although a flower may seem to appear overnight, its **petals** take several days to form inside a bud. Finally, the sepals, or thin green covers on the outside of a bud, open. The beautiful flower petals spread out quickly and begin to grow.

You can see both closed and opening buds in this picture of rose flowers.

Some plants, such as lupine, have several small flowers on a long stem. The bottom flowers generally open before the ones above them.

12. What is a flower's job?

Flowers have the important job of making seeds. A plant's flower must be **fertilized** in order to make seeds. Birds and bugs often help fertilize plants. Many flowers turn into fruits, such as apples, oranges, cherries, strawberries, and tomatoes, which have one or more seeds.

After several months, these flowering trees will produce peaches. Peaches have one big seed, called a stone or a pit.

13. Does every plant start from a seed?

No. Some plants begin in other ways. A green, leafy plant called a fern makes tiny **spores** under its leaves. After the spores are ready, they drop from the leaves and start new fern plants in the dirt below.

Some plants, such as potatoes, can grow new plants from a tuber. Tubers are the bumpy underground stems of certain plants.

The brown dots on the underside of this common polypody's leaf are spore clusters, or places where the fern's spores form.

Cyclamen is another plant that grows from a tuber. These plants grow wild in parts of Europe and the Middle East. In the United States, they are often grown in pots indoors.

Tubers have dark spots called eyes. New stems and leaves grow from the eyes.

Strawberry plants can send out runners to create new plants. Runners are long, thin stems that spread out on the ground and grow new plants from their buds.

Both wild strawberries, such as this one, and strawberries that grow on farms can send out runners to make new plants.

14. Do seeds ever move far from the parent plant?

Many seeds have their own special way of traveling. Some plants have seeds called burrs that catch onto fur or clothes. Burrs ride on animals and people to new places. Dandelion seeds are light and have a fluffy, white top. A puff of wind will blow the seeds far away. Maple-tree seeds have two fans that cause them to spin and drift away as they fall.

When you blow on a fuzzy dandelion, you are helping the plant spread its seeds.

Milkweed plants have pods, or parts that hold their seeds. In the fall, these pods break open to let the seeds escape. Each seed has fine, feathery threads that let it float on the wind.

Some seeds are in fruits. When animals eat fruit, they swallow the seeds inside. Later, animals move on to new places and pass these seeds in their waste.

This cedar waxwing is eating berries from a plant called serviceberry. Cedar waxwings are important seed spreaders, as they eat mostly berries and other fruit.

15. What makes a seed start to grow?

Seeds need air, water, and the right **temperature** to grow.

Farmers plant corn in the early spring. It generally takes corn between 4 and 12 days to start growing.

16. How quickly do seeds start to grow?

Some seeds start growing almost at once. However, most seeds lie in the ground until the time is right for them to sprout, or start growing. In many places, seeds drop from plants in the fall but do not start growing until spring's warm, wet weather comes.

Desert wildflowers, such as this desert lupine, often wait for a rainy year to sprout. Some desert lupine can wait as long as 10 years to sprout.

17. Why else might a seed take time to sprout?

Certain kinds of plants, such as jack pines, need fire to sprout. The pinecones that hold these trees' seeds open only when they are burned. Plants that live in very dry places often wait several years for a wet season to come before they sprout. In 2005, **scientists** even got a 2,000-year-old date palm seed to sprout!

As jack pines do, lodgepole pines need the heat of a fire to open their pinecones and let their seeds out. These trees are common in western North America, where there are a lot of wildfires.

18. Do plants ever eat meat?

The Venus flytrap, pitcher plant, and sundew all eat meat in the form of bugs.

This Venus flytrap is eating bugs. Though people now grow Venus flytraps all over the world, they first came from the wetlands of North Carolina and South Carolina.

The Venus flytrap has leaves that look like a mouth with teeth. When a bug lands inside these leaves, they suddenly snap shut! This plant uses a powerful liquid called acid to turn a bug into juice. The leaves on a pitcher plant are shaped like cups. These cups have red lips that draw bugs near. If a bug goes inside the cup, it drowns and is **digested** by liquid at the bottom. The sundew has leaves with sticky red fingers that trap bugs. The long leaf will curl around its catch and digest it.

The cups of some pitcher plants, such as this one, hang from stems. Other pitcher plants have cups that rise up directly from the ground. There are more than 100 kinds of pitcher plants.

21

20. How many different kinds of plants are found in the world?

No one knows the exact number of different plants on Earth, but scientists think there are more than 400,000 kinds of plants living today. Not all plants grow on land. Some, such as reeds and water lilies, live in the water. Scientists believe there are some kinds of plants that no one has yet found. Perhaps you will become a plant scientist someday and discover a new type of plant!

The world is full of interesting plants. These trees are baobabs from the African country of Madagascar. The trees have huge, wide trunks and can be grow to be 82 feet (25 m) tall!

Glossary

carbon dioxide (KAR-bin dy-OK-syd) A gas that plants take in from the air and use to make food.

chlorophyll (KLOR-uh-fil) Green matter inside plants that allows them to use energy from sunlight to make their own food.

digested (dy-JEST-ed) Broke down food.

evergreen (EH-ver-green) A shrub or tree that has green leaves or needles all year long.

fertilized (FUR-tuh-lyzd) Put together male cells and female cells to make seeds.

glucose (GLOO-kohs) A sugar that plants make to use as food.

minerals (MIN-rulz) Natural things that are not animals, plants, or other living things.

oxygen (OK-sih-jen) A gas that has no color or taste and is necessary for people and animals to breathe.

petals (PEH-tulz) Parts of a flower.

photosynthesis (foh-toh-SIN-thuh-sus) The way in which green plants make their own food from sunlight, water, and a gas called carbon dioxide.

scientists (SY-un-tists) People who study the world.

spores (SPORZ) Special cells that can grow into new living things.

temperature (TEM-pur-cher) How hot or cold something is.

Index

A
air, 4, 9, 18

B
bug(s), 13, 20–21

C
chlorophyll, 8–9

F
flower(s), 10–13

fruit(s), 4, 13, 16

M
minerals, 6, 11

N
nuts, 4

O
oxygen, 4, 9

P
petals, 12
photosynthesis, 8–9

S
spores, 14

T
temperature, 18
trees, 4, 9

Web Sites

Due to the changing nature of Internet links, PowerKids Press has developed an online list of Web sites related to the subject of this book. This site is updated regularly. Please use this link to access the list:

www.powerkidslinks.com/20sci/plant/